ALBERT EINSTEIN

The Genius who Changed the Face of Physics

Written by Julie Lorang
Translated by Rebecca Neal

History   50MINUTES.com

ALBERT EINSTEIN

KEY INFORMATION

- **Born:** 14 March 1879 in Ulm, Germany.
- **Died:** 18 April 1955 in Princeton, USA.
- **Famous discoveries:** the theory of special relativity (1905) and the theory of general relativity (1915) with the formula $E = mc^2$.
- **Impact of his discoveries:**
  - Einstein profoundly altered the way we look at the concepts of space and time (by introducing the idea of black holes, the fourth dimension, etc.);
  - He played a role in the development of nuclear fission and informed President Franklin D. Roosevelt of the possibility of creating an atomic bomb;
  - He revolutionised physics by opening up new fields of study, namely nuclear physics and particle physics.

INTRODUCTION

Even today, Albert Einstein is a household name across the world: he has become the ultimate symbol not only of genius, but also of the humorous and eccentric scientist. How did this nomadic physicist, who left us the celebrated formula $E = mc^2$, become one of the most important historical figures of the 20th century? In order to comprehend this spectacular personality, we must plunge back into Einstein's troubled century and understand how his strikingly modern scientific and political ideas turned our ideas and way of seeing the world upside down.

Einstein was a brilliant scientist, and he shattered the traditional models inherited from the most eminent figures in physics, from Aristotle (384-322 BC) to Galileo (1564-1642) and Isaac Newton (1642-1727). His theories of special relativity and general relativity revolutionised the concepts of space and time and laid the foundations for modern physics. He was awarded the 1921 Nobel Prize in Physics for his explanation of the photoelectric effect, a topic deemed less subversive than the theory of relativity, which was the subject of heated debate at the time the prize was awarded.

However, Einstein's rise to worldwide fame was also due to his identity and philosophical and political commitments. As a Jew, German, pacifist, Communist and Zionist, he lived through the major historical events of the 20th century, and his clearly stated stances made him a particularly modern figure.

A BRIEF HISTORY OF PHYSICS

Over the centuries, the greatest thinkers and scientists, like Einstein, have tried to unravel the mysteries of the universe. They have all offered very different responses to the most important natural phenomena.

ARISTOTLE AND THE NATURAL SCIENCES

Aristotle was one of the most important and most famous intellectuals in ancient Greece. He was the father of natural sciences (*physis* means 'nature' in Greek, a concept which at that time encompassed biology and physics), as well as metaphysics, the branch of philosophy that seeks the reasons behind the existence of our universe. He formulated concepts based on the principles of logic and in particular of syllogism, which involves logical reasoning comprising two propositions which lead to a logical conclusion.

AN EXAMPLE OF A SYLLOGISM

A famous example of a syllogism goes as follows: all men are mortal; Socrates is a man; therefore, Socrates is mortal. For the syllogism to be correct, it must remain logical.

According to Aristotle, all living beings are made up of the four primary elements (earth, air, water and fire), while heavy bodies, such as stars and planets, are made up of aether, a divine substance. Aristotle also formulated the geocentric model: he placed the Earth, which he viewed as immobile, stationary and immovable, at the centre of the universe, with the stars moving around it in perfectly circular orbits. This model was taken up by the Church, because it put man, and therefore divine creation, at the centre of the universe.

Aristotle also thought that there was a clear separation between the Earth and the sky, with each one acting differently depending on the nature of the bodies that it was made up of: heavy bodies such as earth and water naturally moved towards the centre of the planet because of their mass, while light bodies such as fire and air were destined to move towards the sky.

GALILEO AND HELIOCENTRISM

Galileo challenged some Aristotelian concepts and paved the way for a more scientific physics. Through an observation of nature rather than just logic, he disproved Aristotle's claim that bodies fall towards the surface of the Earth more or less quickly depending on their mass. According to legend, Galileo brought the professors of the University of Pisa together at the foot of the city's famous tower and dropped two objects with different masses which reached the ground at the same time, thus confirming his theory.

In spite of the importance of this discovery, Galileo is famous above all for his rejection of Aristotle's idea of geocentrism, supported by the Catholic Church, in favour of heliocentrism. With the development of the telescope, which allowed closer observation of the cosmos, Galileo noted that the Milky Way is made up of many stars and that some planets, such as Jupiter, are at the centre of miniature systems, a truth that he put forward after drawing a parallel between the satellites which orbit a planet and our solar system. Even more importantly, he observed that Earth was not at the centre of the system, but that it orbited the sun, which came as a great shock to people at the time, as their ideas were largely shaped by religion.

GALILEO AND PERSECUTION BY THE CHURCH

Galileo had some disagreements with the Church because of his theory of heliocentrism, which placed the sun at the centre of the universe. The Inquisition also ordered him to stop teaching this theory, before forcing him to describe it as just a hypothesis alongside the Aristotelian model. However, Galileo disobeyed these orders and managed to prove the movement of the Earth. Faced with the threat of the stake, he probably never uttered the famous words *"Eppur si muove"* ("And yet it moves"), but he was still placed under house arrest. Galileo was only officially exonerated by the Church in 1992, while John Paul II (1920-2005) was pope.

NEWTON AND GRAVITATION

Even today, Isaac Newton remains one of the greatest scientists in history thanks to his discovery of the principle of gravitation which, according to legend, came to him when he saw a falling apple. By developing differential and integral calculus, which he needed to confirm his theory, Newton proved that there is a gravitational force which attracts all bodies in proportion to their mass and in inverse-square proportion to the distance between them. The falling of bodies is therefore due to a force rather than to their intrinsic nature, as Aristotle thought, and this force can be calculated using the following formula:

$$F = G \, \frac{m1 \times m2}{d^2}$$

This law expresses, in newtons, the fact that two bodies with masses m_1 and m_2 are attracted to each other with a force, F, which is proportional to the product of their masses and inversely proportional to the square of the distance between them, d. The letter G represents the gravitational constant, equivalent to 6.667×10^{-11}.

From 1684 onwards, Newton used the principle of gravitation for questions of astronomy and calculated the elliptical (not spherical) trajectory of the planets around their sun. In his *Philosophiæ Naturalis Principia Mathematica* (*Mathematical Principles of Natural Philosophy*, 1687), he introduced several laws that were fundamental for the subsequent development of physics. Among other ideas, he set out the principle of inertia, stating that the natural state of a body is at rest, and explained some movements, such as the reaction of a body to an external force. Newtonian physics was of vital importance, because it enabled the explanation of phenomena that were still mysterious at the time, such as the movement of the planets and their satellites, tides, and so on.

EINSTEIN'S LIFE

Einstein photographed while delivering a lecture in Vienna, 1921.

A REBELLIOUS STUDENT

Einstein was born on 14 March 1879 to a middle-class family in the small Bavarian city of Ulm. His parents, Hermann Einstein (1847-1902), who ran a business selling electronic equipment, and Pauline Koch (1858-1920) were both of Jewish origin, but opted for a secular education to allow their son Albert and their daughter Maria (1881-1951) to advance in German society.

The boy who would later be considered a genius did not particularly shine at school to begin with, and his strong personality resulted in a series of disciplinary problems in the very strict and authoritarian Prussian schooling system. Einstein struggled to find his place in the Germany of that time, which was nationalistic and already in the grip of a latent anti-Semitism. When his parents decided to move to Italy for business in 1894, the teenaged Albert saw an opportunity to escape an environment he hated: he left school in the middle of the academic year and gave up his German citizenship to become stateless.

After an initial failure, he finally completed his schooling in Switzerland, where the education system gave him more freedom, before enrolling at the prestigious Federal Polytechnic School in Zürich in 1896. Although his professors viewed him as headstrong, his intelligence and the ease with which he learned allowed him to shine. While studying there, he met his future wife Mileva Maric (1875-1948), an Orthodox Serb who was also studying physics.

THE DIFFICULT YEARS

In 1900, Einstein finished his degree in physics with outstanding results, but the professors at the Polytechnic did not trust him because of his personality, and he was turned down for all the teaching posts he applied to. The young graduate acquired Swiss citizenship in 1901 and supported himself during this time by giving private lessons.

Einstein also faced personal difficulties in the early 20th century. Mileva fell pregnant before the two were married, and returned to her parents in Serbia to have the baby. She gave birth to a daughter named Lieserl in 1902, but all traces of the little girl vanished shortly afterwards. The existence of this child, mentioned in Mileva's letters, only came to light after Einstein's death and gave rise to some outlandish rumours. Was Lieserl given up for adoption or did she die shortly after birth? The mystery of the Einsteins' first child has still not been resolved.

In 1904, things finally began to look up when Einstein obtained a job as an expert at the patent office in Bern, thanks to his loyal friend Marcel Grossmann (Hungarian mathematician, 1878-1936). The same year, he married Mileva, against his parents' wishes. The couple went on to have two children, Hans Albert (1904-1973) and Eduard (1910-1965), who suffered from schizophrenia and sadly died in a psychiatric hospital.

1905: EINSTEIN'S MIRACLE YEAR

1905 was a year of intense intellectual activity and recognition for Einstein. In the space of just a few months, he wrote four papers which would be published in the journal *Annalen der Physik*. These papers included the theory of special relativity and his study of the photoelectric effect, both of which would revolutionise physics and our understanding of the world.

His work made an impression in the great universities and allowed him to fully embrace a career as a physicist: he received his doctorate from the University of Zürich in 1906, and was treated like royalty across the world.

EINSTEIN THE SUPERSTAR

This was the beginning of a new chapter for Einstein, as he became a leading light in the scientific world as well as a popular celebrity. In 1911, he represented the Austro-Hungarian Empire at the first Solvay Conference in Brussels, where he rubbed shoulders with the leading physicists of the time.

Einstein (standing, second from right) at the first Solvay Conference in 1911, photograph taken by Benjamin Couprie.

Subsequently, he was offered jobs by a number of prestigious universities. As such, he taught at the German Charles-Ferdinand University in Prague in 1911, before getting his own back and becoming a professor at the Federal Polytechnic School in Zürich. The following year, he returned to his native country and settled in Berlin, where he was appointed director of the Kaiser Wilhelm Institute for Physics.

His career and his many travels got in the way of his marriage to Mileva, and the two separated in 1914. Five years later, he married his cousin and mistress Elsa Löwenthal (1876-1936).

Einstein continued his research and completed his theory of special relativity by setting out his theory of general relativity. In 1921, his work on the photoelectric effect was recognised with the ultimate reward: the Nobel Prize in Physics.

THE RISE OF NAZISM AND HIS FLIGHT TO AMERICA

The years following the First World War (1914-1918) saw the spread of anti-Semitism in Germany, and once again Einstein suffered because of his origins. This rejection led the scientist to grow closer to the Jewish community, which he was previously not very familiar with, and to embark on his first political activity by campaigning for the creation of a Jewish homeland in Palestine. In 1921, he travelled around the world to raise funds for the creation of the Hebrew University of Jerusalem. He sat on its Board of Governors when it opened in 1925.

When the Nazi Party, led by Adolf Hitler (1889-1945), came to power in 1933, Einstein became one of the regime's priority targets and was forced to flee Germany. He first moved to De Haan in Belgium, before definitively leaving Europe for the USA, where he taught at Princeton University. Terrified by the events that were shaking Europe, he wrote a letter to President Franklin D. Roosevelt (1882-1945) urging him to begin research on the atomic bomb. This letter went on to have a catastrophic impact on the course of history, as it helped lead to the bombings of Hiroshima and Nagasaki on 6 and 9 August 1945.

Atomic bomb exploding above Nagasaki, 9 August 1945.

After becoming an American citizen in 1940, Einstein spent the rest of his life campaigning in favour of pacifism. In spite of his political commitment, he turned down the offer to become President of Israel in 1952, shortly after the country's foundation.

He died in Princeton Hospital on 18 April 1955 following the rupture of an abdominal aortic aneurism. He was cremated and his ashes were scattered in Delaware, on the East Coast of the United States.

When Einstein died, the forensic pathologist Thomas Harvey secretly removed his brain in order to study it and try to unravel the mystery of his exceptional intelligence. The incident came out in the press some 20 years later and caused a great scandal. However, it was not until several years later that Harvey agreed to release his results and suggested that other scientists continue his studies. In the end, it turned out that there was nothing out of the ordinary about Einstein's brain, suggesting that his brilliance could instead stem from his curiosity.

EINSTEIN'S GREATEST RESEARCH AND STRUGGLES

THE FATHER OF MODERN PHYSICS

Einstein revolutionised modern physics with his theories of special relativity and general relativity. Through these ideas, he paved the way for physics in the 20th century and for a vibrant and surprising world.

The theory of special relativity

In 1905, Einstein published four revolutionary articles in the journal *Annalen der Physik*. The most important of these papers concerned the theory of special relativity, which turned everything we thought we knew about the universe on its head.

The fundamental premise of the theory of relativity is that the laws of physics are the same for all observers in uniform motion, whatever speed they are travelling at. For example, if a ping-pong table were placed in a moving train, observers inside the train and on the platform would see very different things: a person on the platform would see the ping-pong ball bouncing while travelling tens of metres at high speed, whereas a person standing in the carriage by the table would only see it moving a few centimetres at a much lower speed. These two observers would therefore witness very different realities, which can only be explained by the relativity of their experience.

However, Einstein went even further by revolutionising the concepts of absolute space and time, thus challenging the idea of a unique frame of reference that is identical for every observer.

Einstein stated that time is intrinsically linked to space to form what is called space-time, a matrix which can be distorted and which makes up the universe. This notion is not easy to grasp, as it seems to contradict the evidence of our senses. It also took several years for some physicists to start supporting it. Einstein offered a humorous illustration of this theory: "Put your hand on a hot stove for a minute, and it seems like an hour. Sit with a pretty girl for an hour, and it seems like a minute. That's relativity."

As part of his theory on special relativity, he came up with the famous equation $E = mc^2$, where E represents energy, m represents mass and c represents the speed of light (3.00×10^8 m/s). In this way, he proved that splitting an atom results in much more energy. This equation is used in particular to calculate the amount of energy produced when a quantity of matter is converted into electromagnetic radiation. It was therefore used to develop nuclear fission, which led to the creation of the atomic bomb. Because of the high speed of light, the effect of mass converted into energy is immense: the bomb which destroyed Hiroshima only weighed a few dozen grams, but claimed the lives of 100 000 people. However, this is not its only use: it also enabled the development of PET scans, in which radiation is used to provide an image of the inside of the body.

The theory of general relativity

In 1915, almost ten years after setting out his theory of special relativity, Einstein completed it by incorporating the notion of gravity. According to him, gravity was very different from the explanation given by Newton. Newton believed that it was the gravitational force that attracted bodies to one another and made apples fall. However, this did not explain how one body was attracted to another. Furthermore, Newton imagined that the two bodies were moving in a void, and consequently was unable to resolve the nature of space.

In 1865, the theory of electromagnetism formulated by James Clerk Maxwell (Scottish physicist, 1831-1879) provided some partial answers to the many questions left by Newton's theory. Maxwell put forward the hypothesis that the propagation of light, which happens at a constant speed, is possible thanks to electric and magnetic waves, which make up an electromagnetic field. Maxwell discovered the existence of this field and imagined that a kind of immaterial fluid known as luminiferous aether fills space and constitutes the necessary medium for the propagation of electromagnetic waves.

Einstein was fascinated by electricity from childhood and soon took an interest in the theory of the electromagnetic field as set out by Maxwell. He was convinced that there was also a gravitational field and, after several years of research, he put forward the hypothesis that this gravitational field is not spread out in space, but rather constitutes space itself. In other words, Einstein synthesised the Newtonian conception of space, in which bodies moved, and Maxwell's field made up of waves.

This discovery was important because it revolutionised our ideas about time and space, and because it provided a fairly simple explanation for a number of natural phenomena which remained mysterious. According to Einstein, space-time is a flexible material component: it can wave, expand and twist. This idea allows us to understand, for example, why the Earth is attracted towards the sun: this enormous star curves space, and our planet simply follows the slope, like a marble in a funnel. This curving explains why planets orbit around a star.

The structure of space and the Earth's orbit around the sun

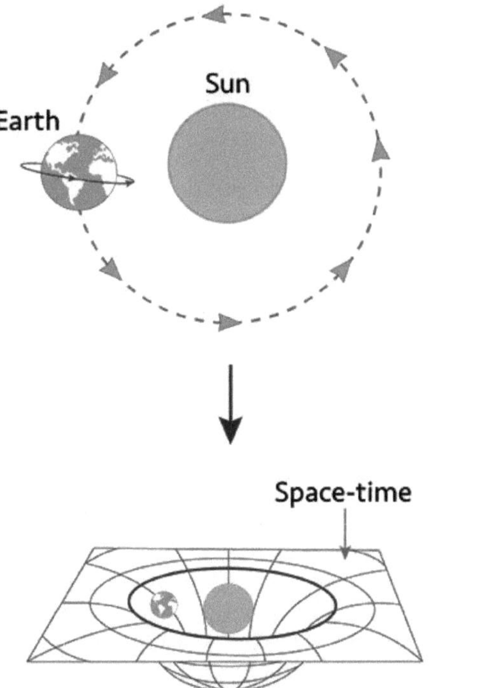

Einstein's dismantling of the traditional frame of reference of space-time quickly made him famous not only among his peers, but also with the general public.

Several artists were influenced by this deconstruction and tackled space and time in their work. For example, Pablo Picasso (1881-1973) deconstructed the pictorial space with cubism, the Austrian composer Arnold Schoenberg (1874-1951) worked on time and harmony in music, and the French psychologist and philosopher Henri Bergson (1859-1941) put forward the idea of the subjectivity of duration.

The wave and photon theory of light

Although Einstein's work on the photon theory of light is less well-known than his work on special relativity and general relativity, it nonetheless earned him the highest honour for a scientist, the Nobel Prize in Physics, in 1921.

In 1905, Einstein became interested in a debate which had divided the scientific community for decades: what is the nature of light? Is it made up of waves or particles? The two sides based their ideas on the work of two great scientists from the 17th century, the Dutchman Christiaan Huygens (1629-1695) and the Englishman Isaac Newton. Huygens put forward a wave-based theory of light, whereas Newton believed that it was made up of small particles. The limitations of Huygens' research and Newton's great prestige meant that the second theory was accepted, and it was not questioned for two centuries.

However, during the 19th century a number of physicists, including the Frenchman Augustin Fresnel (1788-1827), the founder of modern optical science, and the British scientist

Thomas Young (1773-1829), returned to the question. Their experiments proved that when light is diffracted, it forms an interference pattern comparable to waves in water. In other words, Fresnel and Young agreed with Huygens' theory of waves. In 1850, the physicist Léon Foucault (1819-1868) even managed to calculate the speed of propagation of the waves which make up light. The wave theory of light then had the upper hand, at least until Einstein's arrival.

In 1905, Einstein surprised scientists by stating that neither Huygens nor Newton were really wrong. He explained that light has a dual nature: it is at once wave-based and made up of small particles of light energy. He called these particles "light quanta", but today they are more commonly known as photons. He also developed the Planck-Einstein relation, which calculates the energy of photons, E, by multiplying the frequency of light, v (the letter "nu" in the Greek alphabet), by the Planck constant h (6.626×10^{-34}):

$$E = h \cdot v$$

This particle-wave duality is now widely recognised by scientists across the world.

EINSTEIN'S SOCIAL AND POLITICAL COMMITMENT

Already in the 16th century, the French humanist François Rabelais (1483-1553) had written that "science without conscience is the ruin of the soul". This maxim, which is more relevant now than ever, provides a good summary of Einstein's career. Indeed, he did not merely observe and conceptualise physics problems; he also closely followed the political problems of his time and campaigned for several causes close to his heart.

Einstein and Germany

Einstein's identity and personal history were closely intertwined with the major issues of the 20th century. As a German and nonobservant Jew, the young Albert endured rejection from a very young age in a society gripped by nationalism. At the age of 16, he renounced his German citizenship because he was strongly opposed to the political situation in his country. He nonetheless remained attached to his homeland, and his post as director of the Kaiser Wilhelm Institute for Physics allowed him to take his personal revenge.

When the First World War broke out, Einstein was horrified to realise that nationalism and hatred were not the prerogative of the popular classes, but that many of his colleagues and students also threw themselves into this mass hysteria. For example, when Germany invaded neutral Belgium and committed many acts of violence which were met with widespread criticism, 93 German intellectuals signed a manifesto in which they affirmed their support for Kaiser Wilhelm II (1859-1941). As a staunch pacifist, Einstein categorically refused to sign this manifesto, a decision which was viewed as treasonous by many Germans.

After the war, he became a member of the German League

for Human Rights and worked to bring France and Germany closer. However, the rise of Nazism brought discussions between the two countries to an abrupt end. When Hitler came to power, Einstein was persecuted, his work was branded "Jewish physics", and leading scientists called for his theories to be removed from physics textbooks. The situation continued to worsen until he was forced to flee his country.

Einstein would never forgive Germany for the Holocaust and the extermination of six million Jews, and did not want to be associated with the country for the rest of his life. However, he remained strongly influenced by German culture: he never attained complete mastery of English, and never really got used to Anglo-Saxon culture.

A moderate Zionist

Nationalism and the bitter taste left by the defeat in 1918 led Germany to seek a scapegoat in the Jews. This period also saw a major influx of Jews from Eastern Europe, who had been driven out by pogroms. They were given a hostile reception in Germany, as well as in some other countries. In light of this situation, Einstein thought it essential to create a haven for the Jews and fully committed himself to this project. For example, he took an active part in the foundation of the Hebrew University of Jerusalem by raising funds in the USA.

However, Einstein was far from a radical Zionist, and did not think that Israel should belong to the Jews to the detriment of the Palestinians. He strongly emphasised the need to open a dialogue between the two countries in order to find a solution that would allow them to live in harmony. Einstein was therefore in favour of creating a haven for the Jews, alongside the Palestinians, but did not support the establishment of a Jewish state.

EINSTEIN, PRESIDENT OF ISRAEL?

When Chaim Weizmann, the first President of Israel, died in 1952, the position was offered to Einstein, who turned it down, claiming that he lacked the requisite political knowledge and experience. On the subject, he said: "Equations are more important to me, because politics is for the present, but an equation is something for eternity" (Hawking, 2009).

The atomic bomb

Einstein is often, wrongly, considered to be the father of the atomic bomb. Although he did play a role in the emergence of this weapon of mass destruction, he did not create it.

When Hitler came to power, many European scientists were forced to flee the continent and seek refuge in the USA. This community, which was traumatised by the horrors of Nazism, was afraid that Germany would develop a weapon of mass destruction, the atomic bomb. Two physicists, Einstein and Leo Szilard (1898-1964), then wrote a letter urging President Franklin D. Roosevelt to get there before the Nazis and carry out research on nuclear fission. This gave rise to the Manhattan Project, classified as a military secret, in 1942. Einstein did not take part in the project, but his formula $E = mc^2$ was used extensively to turn mass into energy. Before too long, the American researchers managed to make an atomic bomb. Two bombs were then dropped on Hiroshima and Nagasaki, thus putting an end to the Second World War.

Einstein never forgave himself for his role in creating such a monstrous invention, and spent the rest of his life campaigning for the establishment of an international body to control nuclear energy. The Treaty on the Non-Proliferation of Nuclear Weapons was signed by 189 countries in 1968 and came into force in 1970, but Einstein did not live to see it.

IMPACT

NEW PERSPECTIVES IN PHYSICS

The greatest impact of Einstein's work is the development of a kind of physics that is so modern that some of his hypotheses have only just been confirmed. Several of his theories, such as the existence of black holes and the creation of the universe through the Big Bang, seemed so outlandish to scientists at the time that they were not taken seriously when they were first put forward. They could only be tested and proved once new scientific techniques had been developed. This moved them from the realm of science fiction to the realm of scientific reality.

For example, during his lifetime Einstein had stated that black holes exist and developed a theory that many viewed as borderline esoteric. He claimed that when a star dies, what remained of it is no longer supported by heat, so it collapses in on itself under its own weight, causing space to curve and creating a hole. It was not until 1971 that a NASA satellite captured Cygnus X-1 x-ray emissions, thus proving that black holes exist.

Likewise, Einstein's equations led to confirmation that the universe is expanding and that only an explosion could have given it this momentum. This was the beginning of the Big Bang theory, which was proposed by the Belgian astronomer and physicist Georges Lemaître in 1931.

That is not all. On 14 September 2015, American scientists

managed to prove the existence of gravitational waves, which had been predicted almost a century earlier by Einstein's theory of general relativity. He had stated that the universe was made up of a matrix, space-time, which could become distorted and produce vibrations, known as gravitational waves. These waves, which were theorised in 1916, had never been directly detected until that day in September 2015, when two inferometers (devices which measure distances using the interference of light) situated at opposite ends of the USA finally captured them following the collision of two black holes light years away from Earth. To do this, scientists had had to wait for decades in order to develop sensors sophisticated enough to intercept gravitational waves, which can be confused with certain seismic movements on Earth. This incredible discovery once again proved that Einstein's theories were valid and opened up new perspectives for astrophysicists. Until now, astronomers have only employed light, by using wavelengths along the spectrum, to understand the different phenomena of the universe. Thanks to gravitational waves, in the near future it will be possible to explore phenomena that lie beyond the power of light, such as the core of a massive star at the end of its life, for example. There is no doubt that gravitational waves will continue to cause a stir for many years to come.

EINSTEIN'S CONTINUED UBIQUITY

Even today, Einstein remains omnipresent in the field of science. For example, an einstein is a unit of measurement which designates the energy of one mole of photons, the particle whose existence he predicted in

1905. He also gave his name to einsteinium, a white radioactive metal and one of the elements of the periodic table. Finally, every year on 14 March, Einstein's birthday, scientists across the world celebrate Pi Day, as 3, 1 and 4 are the first significant digits of π. Although there is no link between this number and Einstein's work, the choice of his birthday shows his importance for the scientific community.

SUMMARY

- Albert Einstein was born in Ulm, Germany in 1879 to Jewish parents. From a very early age, he rebelled against the nationalism and anti-Semitism of his native country. He refused to accept the measures taken by Germany and renounced his citizenship at the age of 16 to become stateless. After studying in Switzerland, he obtained Swiss citizenship in 1901, before becoming an American citizen in 1940.

- In 1896, he began studying physics at the Federal Polytechnic School in Zürich, where he met his future wife Mileva Maric. They married in 1904 and had three children together, including a little girl who mysteriously disappeared in Serbia.

- 1905 was Einstein's *annus mirabilis*: he wrote four papers containing innovative theories for the journal *Annalen der Physik*. The theory of special relativity shattered the idea of a unique reference point for space and time. While researching the subject, he developed his famous equation $E = mc^2$.

- These papers earned him the recognition of his peers. He travelled all over the world and taught first at the German Charles-Ferdinand University in Prague, then in Zürich, and finally in Berlin.

- In 1915, he completed his theory of general relativity with his theory of special relativity, by integrating gravity into it. In a way, he fused Isaac Newton's theory of relativity and James Clerk Maxwell's theory of electromagnetism: the universe and the electromagnetic field were a single entity, and this entity could expand and become deformed, with a similar impact on space-time.

- Although he was not observant, Einstein was of Jewish origin, and he put a great deal of effort into the creation of the Hebrew University of Jerusalem, which was founded in 1918. He also campaigned for the creation of Israel, but warned against the excesses of a state built to the detriment of the Palestinian people. He also turned down the opportunity to become president of the young state in 1952.
- In 1921, he received the highest reward for scientists, the Nobel Prize in Physics, for his work on the photoelectric effect, which was viewed as less subversive than the theory of relativity.
- In 1933, he left Germany because of the Nazi threat and moved to Belgium, then Princeton, USA. Traumatised by the events that were shaking Europe, he wrote a letter to President Franklin D. Roosevelt urging him to develop nuclear fission. The Manhattan Project was established shortly afterwards and culminated in the bombings of Hiroshima and Nagasaki in 1945.
- Einstein died in Princeton in 1955.

FIND OUT MORE

BIBLIOGRAPHY

* Galison, P. (2003) *Einstein's Clocks and Poincaré's Maps: Empires of Time*. New York: W.W. Norton & Company, Inc.
* Hawking, S. (2009) *A Brief History of Time: From Big Bang to Black Holes*. New York: Random House.
* Lacayo, R. (2011) *Albert Einstein: The Enduring Legacy of a Modern Genius*. New York: Time to Explore.
* Le Soir Mag (2015) *L'extraordinaire saga du cerveau d'Einstein*. [Online]. [Accessed 9 March 2017]. Available from: <http://www.lesoir.be/1060561/article/soirmag/soirmag-histoire/2015-12-03/l-extraordinaire-saga-du-cerveau-d-einstein>
* Maier, C. and Simon, A. (2016) *Einstein*. London: Nobrow Ltd.
* Rovelli, C. (2016) *Seven Brief Lessons on Physics*. London: Penguin.

ADDITIONAL SOURCES

* Bodanis, D. (2016) *Einstein's Greatest Mistake: The Life of a Flawed Genius*. London: Little, Brown Book Group.
* Bodanis, D. (2016) *E=mc²*. London: Pan Books.
* Cox, B. and Forshaw, J. (2010) *Why Does E=mc²?* Philadelphia: Da Capo Press.
* Einstein, A. (2007) *The World As I See It*. New York: Philosophical Library.
* Hawking, S. (2016) *Black Holes: The Reith Lectures*.

London: Bantam.
* Isaacson, W. (2008) *Einstein: His Life and Universe*. London: Simon and Schuster.

ICONOGRAPHIC SOURCES

* Einstein photographed while delivering a lecture in Vienna, 1921. Royalty-free reproduction picture.
* Einstein (standing, second from right) at the Solvay Conferences in 1911, photograph taken by Benjamin Couprie. Royalty-free reproduction picture.
* Atomic bomb exploding above Nagasaki, 9 August 1945. Royalty-free reproduction picture.

History
Business
Coaching
Book Review
Health & Wellbeing